Book Title:

"Sustainable Coffee: How to Choose Eco-Friendly Coffee and Support Sustainable Farming Practices."

By Jenny Koo

Imprint: Independently published.

Copyright © 2024 by Jenny K. Koo. All rights reserved.

No part of this book may be used or reproduced in any manner whatsoever without written permission.

For information, please email to jennykookk@gmail.com

"Sustainable Coffee: How to Choose Eco-Friendly Coffee and Support Sustainable Farming Practices."

By Jenny Koo

Table of Contents

Introduction
- Purpose of the Book
- What Readers Will Learn
- Coffee's Global Significance

Why Sustainability Matters in Coffee
- Environmental and Social Impacts
- Sustainability as a Solution

Overview of Sustainable Coffee Practices
- What is Sustainable Coffee?
- Brief History of Sustainability in Coffee
- Importance of Awareness

Chapter 1: Understanding Sustainable Coffee
- Defining Sustainability in Coffee
- The Journey from Bean to Cup

Chapter 2: Environmental Impact of Coffee Production
- Deforestation and Biodiversity Loss
- Water Usage and Pollution
- Soil Health and Erosion

Chapter 3: Sustainable Farming Practices
- Shade-Grown Coffee
- Organic Farming
- Agroforestry
- Composting and Waste Management

Chapter 4: Certifications and Labels
- Fair Trade
- Rainforest Alliance
- USDA Organic
- Other Certifications (UTZ, Bird Friendly)

Chapter 5: Choosing Eco-Friendly Coffee
- Reading Labels and Certifications
- Supporting Ethical Brands
- Local and Direct Trade Options

Chapter 6: Sustainable Brewing and Consumption
- Eco-Friendly Brewing Methods
- Reducing Waste
- Recycling and Composting

Chapter 7: The Social Impact of Sustainable Coffee
- Empowering Farmers
- Community Development
- Gender Equality

Chapter 8: The Future of Sustainable Coffee
- Innovations in Coffee Production

- Challenges and Opportunities
- Consumer Role

Conclusion
- Recap of Key Points
- Call to Action

Introduction

Purpose of the Book

For many, coffee is more than just a beverage—it's a ritual, a source of comfort, and a vital part of daily life. But have you ever paused to consider where your coffee comes from or the impact it has on the world around you? As the global demand for coffee continues to rise, so too does the importance of making informed choices about the coffee we consume. This book is your guide to navigating the world of sustainable coffee, whether you're a casual drinker or a seasoned aficionado.

The journey from bean to cup is a complex one, involving farmers, traders, roasters, and baristas. Along the way, the choices made by each of these players affect not only the quality of the coffee but also the environment and the lives of those who grow it. By choosing sustainable coffee, you're not only enjoying a better cup but also contributing to the well-being of the planet and the people behind your morning brew.

What Readers Will Learn

In this book, you will explore what it means for coffee to be "sustainable" and how you can make choices that align with your values. You'll learn to decipher the various certifications and labels that adorn coffee packages, understand the environmental and social impacts of coffee production, and discover how your brewing methods can make a difference.

We'll also delve into the future of sustainable coffee, examining innovations and challenges that lie ahead. By the end of this book, you'll be equipped with the knowledge to confidently choose coffee that not only tastes good but also does good.

Coffee's Global Significance

Coffee is one of the most beloved beverages in the world, consumed by millions of people daily. It's also the second most traded commodity globally, right after oil, which highlights its enormous economic significance. But behind this ubiquitous drink is a vast network of farmers, most of whom are small-scale growers in developing countries. These farmers often face challenges such as climate change, unstable market prices, and unfair labor practices.

The environmental impact of coffee production is equally significant. Traditional coffee farming often leads to deforestation, soil degradation, and water pollution. However, sustainable practices can help mitigate these effects, preserving biodiversity and improving soil health. By making conscious choices, you as a consumer can support these positive practices and contribute to a more sustainable future for coffee.

Sustainable coffee is not just a trend; it's a necessity. As you read through this book, you'll discover that the choices you make, from the coffee you buy to how you brew it, have far-reaching impacts. Together, we can ensure that future generations will be able to enjoy coffee as much as we do today.

Why Sustainability Matters in Coffee

Environmental and Social Impacts

Coffee is a global staple, but its production has far-reaching environmental and social consequences. Traditional coffee farming often leads to deforestation, threatening biodiversity and contributing to climate change. Large-scale coffee farms, especially those in sun-exposed monocultures, disrupt ecosystems by removing native trees and plants, which in turn affects wildlife habitats.

Water usage in coffee production is another significant issue. Wet processing methods, common in many coffee-producing countries, use large amounts of water, often leading to contamination of local water supplies with chemicals and organic waste. This pollution harms aquatic life and can make water unsafe for nearby communities.

On the social side, many coffee farmers work under challenging conditions, with low wages, limited access to healthcare, and poor living standards. In many regions, the volatility of coffee prices in global markets exacerbates these issues, leaving farmers vulnerable to poverty and economic instability.

Sustainability as a Solution

Sustainability in coffee production offers a way to mitigate these impacts. By adopting environmentally friendly practices, such as shade-grown coffee, organic farming, and agroforestry, coffee producers can protect biodiversity, improve soil health, and reduce the carbon footprint of their farms. Sustainable practices also promote the efficient use of water and the management of waste, helping to preserve precious natural resources.

On the social front, sustainability ensures that farmers receive fair compensation for their labor, access to education, and better working conditions. Certification programs like Fair Trade and Rainforest Alliance play a crucial role in promoting these standards, ensuring that the benefits of coffee production are shared more equitably among all stakeholders.

Sustainability is not just about protecting the environment; it's about creating a coffee industry that is fair, resilient, and capable of thriving in the long term. By understanding the importance of sustainability, we can make informed choices that support these efforts and contribute to a more sustainable future for coffee.

Overview of Sustainable Coffee Practices

What is Sustainable Coffee?

Sustainable coffee is grown and processed in a way that minimizes environmental impact and promotes social equity. This approach to coffee production focuses on conserving natural resources, protecting ecosystems, and ensuring that farmers are treated fairly and paid appropriately. Sustainable coffee also emphasizes quality, with the understanding that healthy plants, grown in a balanced ecosystem, produce better beans.

Brief History of Sustainability in Coffee

The concept of sustainability in coffee began gaining traction in the late 20th century as the environmental and social impacts of coffee production became more apparent. In the 1980s and 1990s, the rise of organic farming and the introduction of certification programs like Fair Trade marked the beginning of the sustainable coffee movement. These initiatives aimed to address the problems of low wages, poor working conditions, and environmental degradation that were prevalent in many coffee-growing regions.

Over the years, the movement has grown, with more farmers adopting sustainable practices and more consumers seeking out ethically produced coffee. Today, sustainability is a key consideration for many in the coffee industry, from smallholder farmers to large corporations.

Importance of Awareness

Consumer awareness is critical to the success of sustainable coffee. By understanding what sustainability means and why it matters, consumers can make informed choices that support sustainable practices. This includes looking for certifications, buying from ethical brands, and understanding the impact of their coffee consumption.

When consumers demand sustainable coffee, they create a market that values and rewards responsible production. This, in turn, encourages more farmers to adopt sustainable practices, leading to a positive cycle of improvement that benefits everyone involved—from the farmers to the environment, to the end consumer.

Chapter 1: Understanding Sustainable Coffee

Defining Sustainability in Coffee

Sustainability in coffee is a comprehensive concept that integrates environmental protection, social responsibility, and economic viability. It goes beyond the simple notion of being "green" and involves a holistic approach to the entire coffee production process. Sustainable coffee production ensures that the environment is preserved, that farmers and workers are treated fairly and paid equitably, and that the industry remains profitable for everyone involved, from the farmer to the consumer.

Environmental protection is a key component of sustainability. This includes practices such as minimizing the use of synthetic chemicals, conserving water, and maintaining biodiversity by using methods like shade-grown coffee. Shade-grown coffee, for example, allows coffee plants to grow under the canopy of trees, which helps preserve the surrounding ecosystem, supports wildlife, and reduces soil erosion. Organic farming, another aspect of environmental sustainability, focuses on natural methods of pest control and soil fertility, avoiding harmful pesticides and fertilizers.

Social responsibility is equally important in sustainable coffee. It ensures that the people who grow and harvest the coffee are treated with respect and fairness. This means providing fair wages, safe working conditions, and opportunities for education and community development. Social sustainability also includes the empowerment of marginalized groups, such as women and indigenous populations, who often play a crucial role in coffee production but are frequently underrepresented and underpaid.

Economic viability ties the environmental and social aspects together. For coffee production to be truly sustainable, it must also be economically viable. This means that farmers need to earn a living wage that covers the cost of production and provides a profit margin that allows them to reinvest in their farms and improve their quality of life. Economic sustainability also involves creating a transparent and fair supply chain, where the benefits of coffee production are distributed equitably among all participants.

In essence, sustainable coffee is about creating a balance between these three pillars—environmental protection, social responsibility, and economic viability—to ensure that the coffee industry can thrive for generations to come.

The Journey from Bean to Cup

The journey from bean to cup is a complex process that involves several stages, each with its own set of sustainability challenges and opportunities. Understanding this journey helps us appreciate the intricacies of coffee production and the importance of making sustainable choices at every step.

Coffee cultivation typically begins in tropical regions, where the climate and soil conditions are ideal for growing coffee plants. However, traditional coffee farming practices, such as monocropping in full sun, can lead to deforestation, soil degradation, and loss of biodiversity. In contrast, sustainable cultivation practices, like shade-grown coffee, protect the environment by preserving tree cover, maintaining soil health, and providing habitats for wildlife. Organic farming further enhances sustainability by avoiding synthetic chemicals, which can harm the environment and the health of farmers.

Once the coffee cherries are ripe, they are usually harvested by hand. This labor-intensive process ensures that only the best cherries are selected, which is crucial for maintaining the quality of the coffee. However, this stage also highlights the importance of fair labor practices and fair compensation for the workers who perform this demanding work. Ensuring that these workers are paid fairly and treated with respect is an essential aspect of social sustainability.

After harvesting, the cherries are processed to extract the beans. There are several methods of processing, including dry processing, which involves drying the cherries in the sun, and wet processing, which uses water to ferment and wash the beans. Each method has different environmental impacts, particularly concerning water usage and waste management. Sustainable processing practices focus on minimizing water use, reducing waste, and protecting local water resources from contamination.

The green coffee beans are then roasted to develop their flavor. Roasting is a critical step in the coffee production process, and it can be done at various scales, from small artisanal batches to large industrial operations. Sustainable roasting practices include using energy-efficient equipment, minimizing emissions, and ensuring that the roasting process is as eco-friendly as possible.

Finally, the roasted beans are ground and brewed to make the coffee you enjoy. The method you choose for brewing—whether it's a French press, espresso machine, or drip coffee maker—can also impact the sustainability of your coffee habit. Sustainable brewing methods focus on reducing waste, such as using reusable filters and composting coffee grounds, and minimizing energy consumption.

The journey from bean to cup is a testament to the dedication and effort required to produce high-quality coffee. By understanding each stage of this journey, we can better appreciate the importance of sustainability in coffee production and make informed choices that support a more sustainable coffee industry.

Chapter 2: Environmental Impact of Coffee Production

Deforestation and Biodiversity Loss

The environmental impact of coffee production is significant, with deforestation being one of the most pressing issues. Traditional coffee farming often involves clearing vast areas of forest to create space for coffee plantations. This practice, particularly prevalent in sun-grown coffee systems, leads to the destruction of vital habitats for countless species, contributing to a significant loss of biodiversity.

Forests are not only home to diverse wildlife but also play a critical role in regulating the earth's climate by absorbing carbon dioxide. When forests are cleared for coffee plantations, this carbon sequestration capacity is lost, contributing to global warming. Additionally, the removal of trees disrupts local ecosystems, leading to soil erosion and the depletion of nutrients in the soil, which further diminishes the land's productivity over time.

In contrast, sustainable coffee farming practices, such as shade-grown coffee, work to protect these ecosystems. Shade-grown coffee involves cultivating coffee plants under the canopy of existing trees, preserving the forest and its biodiversity. This method not only protects wildlife but also enhances the quality of the coffee, as the shade helps the coffee cherries mature slowly, resulting in a richer flavor profile.

Water Usage and Pollution

Water is essential in coffee production, particularly during the processing stage, where coffee cherries are often washed to remove the pulp before drying. However, the traditional wet processing method uses large amounts of water, often in regions where water is already a scarce resource. This excessive water use can strain local water supplies, leaving communities with limited access to clean water.

Moreover, the wastewater generated during coffee processing is typically loaded with organic material and chemicals, which, if not properly treated, can contaminate local water sources. This pollution can have devastating effects on aquatic ecosystems, killing fish and other wildlife and making the water unsafe for human consumption.

Sustainable coffee practices aim to minimize water usage and manage waste more effectively. Dry processing, for example, uses much less water, though it requires more careful handling to avoid fermentation issues. In areas where wet processing is the norm, some producers have begun using more efficient equipment that recycles water, reducing overall consumption and limiting the impact on local water supplies. Additionally, the use of natural filtration systems, such as constructed wetlands, can help treat wastewater, preventing pollution and protecting local ecosystems.

Soil Health and Erosion

Soil health is another critical environmental concern in coffee production. The intensive farming practices used in conventional coffee production, such as monocropping and the overuse of chemical fertilizers and pesticides, can lead to soil degradation. Over time, these practices strip the soil of essential nutrients, reducing its fertility and making it more susceptible to erosion.

Erosion is a particular problem in coffee-growing regions, which are often located on steep hillsides. Without proper ground cover, rainwater can wash away the topsoil, leading to landslides and further degrading the land's ability to support crops. This loss of topsoil not only reduces coffee yields but also affects the surrounding environment, contributing to sedimentation in rivers and streams, which can harm aquatic life and disrupt water supplies.

Sustainable coffee farming practices address these issues by promoting soil conservation techniques. These include planting cover crops to protect the soil from erosion, using organic fertilizers to maintain soil fertility, and practicing crop rotation to prevent nutrient depletion. By focusing on soil health, sustainable coffee farming not only ensures long-term productivity but also helps protect the broader environment.

Chapter 3: Sustainable Farming Practices

Shade-Grown Coffee

Shade-grown coffee is a traditional farming method that has gained renewed attention for its environmental benefits. In this system, coffee plants are grown under the canopy of taller trees, mimicking the natural conditions in which coffee evolved. The shade provided by these trees helps protect the coffee plants from excessive sunlight, reduces the need for chemical inputs, and supports biodiversity by providing habitats for birds, insects, and other wildlife.

The trees in a shade-grown coffee system also contribute to soil health by dropping leaves that decompose and enrich the soil with organic matter. This natural mulch helps retain moisture, reducing the need for irrigation, and protects the soil from erosion. Furthermore, the diversity of plant species in a shade-grown system can help control pests naturally, reducing the reliance on harmful pesticides.

For consumers, shade-grown coffee often offers a superior flavor profile. The slower maturation of coffee cherries in shaded conditions allows for the development of more complex flavors, resulting in a richer and more nuanced cup of coffee. By choosing shade-grown coffee, consumers can support farming practices that protect the environment and promote biodiversity.

Organic Farming

Organic farming is another cornerstone of sustainable coffee production. Organic coffee is grown without the use of synthetic fertilizers, pesticides, or genetically modified organisms (GMOs). Instead, organic farmers rely on natural methods to maintain soil fertility, control pests, and manage diseases. These methods include composting, crop rotation, and the use of organic fertilizers such as manure and compost.

One of the key benefits of organic farming is its positive impact on soil health. Organic practices help maintain and enhance the natural fertility of the soil, ensuring that it remains productive for future generations. Healthy soil is also more resilient to extreme weather conditions, such as droughts and heavy rains, which are becoming more common due to climate change.

In addition to its environmental benefits, organic farming also has positive social impacts. Organic certification often comes with a premium price, which can provide additional income for farmers. This extra income can be invested in improving living conditions, education, and healthcare for farming communities. By choosing organic coffee, consumers are not only supporting environmentally friendly practices but also contributing to the well-being of coffee farmers and their families.

Agroforestry

Agroforestry is a sustainable farming practice that integrates trees and other vegetation with crops and livestock. In coffee production, agroforestry involves planting coffee alongside various tree species, creating a diversified and resilient ecosystem. This approach offers numerous environmental benefits, including carbon sequestration, improved soil health, and enhanced biodiversity.

In an agroforestry system, trees provide shade for coffee plants, reducing the need for irrigation and protecting the soil from erosion. The diverse plant species in an agroforestry system also support a wide range of wildlife, contributing to the preservation of biodiversity. Additionally, trees can produce other valuable products, such as fruit, nuts, or timber, providing farmers with additional sources of income and reducing their dependence on coffee alone.

Agroforestry is particularly well-suited to smallholder farmers, who often have limited resources and need to make the most of their land. By diversifying their crops, these farmers can reduce their risk and increase their resilience to market fluctuations and climate change. For consumers, supporting coffee produced through agroforestry practices means contributing to a more sustainable and resilient agricultural system.

Composting and Waste Management

Composting is a vital practice in sustainable coffee farming that helps manage waste and improve soil health. Coffee production generates a significant amount of organic waste, including coffee pulp, leaves, and branches. Instead of discarding this waste, farmers can compost it, turning it into a valuable resource that enhances soil fertility and reduces the need for chemical fertilizers.

Composting involves the controlled decomposition of organic matter, creating a nutrient-rich material that can be used to enrich the soil. This process not only reduces the amount of waste generated on the farm but also recycles nutrients back into the soil, promoting healthy plant growth and improving the soil's ability to retain water.

In addition to composting, sustainable waste management practices in coffee production also focus on minimizing the environmental impact of other by-products, such as wastewater from processing. By treating and reusing wastewater, coffee producers can reduce pollution and conserve water resources.

For consumers, choosing coffee from farms that prioritize composting and waste management means supporting practices that reduce environmental harm and promote a circular economy, where waste is minimized, and resources are used more efficiently.

Chapter 4: Certifications and Labels

Fair Trade

Fair Trade certification is one of the most well-known and respected labels in the coffee industry. It was established to address the social and economic challenges faced by coffee farmers, particularly in developing countries. Fair Trade ensures that farmers receive a fair price for their coffee, which covers the cost of sustainable production and provides a living wage.

Fair Trade certification also promotes better working conditions, democratic organization, and community development. It requires that farmers and workers have access to education, healthcare, and safe working environments. The Fair Trade premium, an additional sum of money paid on top of the fair price, is invested in community projects, such as schools, clinics, and infrastructure improvements.

For consumers, choosing Fair Trade coffee is a way to support farmers and their communities, ensuring that the people who grow their coffee are treated fairly and can build a better future for themselves and their families.

Rainforest Alliance

The Rainforest Alliance certification focuses on environmental sustainability, social equity, and economic viability. It ensures that coffee is grown in a way that conserves natural resources, protects wildlife, and promotes the well-being of farming communities.

Rainforest Alliance-certified farms must meet rigorous standards for environmental management, including the conservation of biodiversity, soil health, and water resources.

In addition to environmental criteria, Rainforest Alliance certification also addresses social and economic issues. It requires that workers are treated fairly, with access to decent wages, safe working conditions, and the right to organize. The certification also promotes community development, encouraging farmers to invest in education, healthcare, and other social services.

For consumers, Rainforest Alliance certification provides assurance that their coffee is produced in a way that respects both people and the planet. By choosing Rainforest Alliance-certified coffee, consumers can support sustainable farming practices that contribute to a healthier and more equitable world.

USDA Organic

The USDA Organic label is a widely recognized certification that indicates coffee has been produced without the use of synthetic fertilizers, pesticides, or genetically modified organisms (GMOs). Organic coffee farming focuses on natural methods of pest control, soil fertility, and crop rotation, which help maintain the health of the environment and produce high-quality coffee.

USDA Organic certification requires that coffee farms adhere to strict standards for organic production, including the use of organic seeds, the management of soil health, and the prevention of contamination from non-organic sources. The certification process also includes regular inspections to ensure compliance with organic standards.

For consumers, the USDA Organic label provides confidence that their coffee is produced in an environmentally friendly way, without harmful chemicals or GMOs. By choosing organic coffee, consumers can support farming practices that protect the environment and promote sustainability.

Other Certifications (UTZ, Bird Friendly)

In addition to Fair Trade, Rainforest Alliance, and USDA Organic, there are several other certifications that promote sustainable coffee production. UTZ certification, for example, focuses on sustainable farming practices, traceability, and social responsibility. UTZ-certified farms must meet standards for environmental management, worker rights, and community development, similar to those of other certifications.

Bird Friendly certification, developed by the Smithsonian Migratory Bird Center, specifically focuses on protecting bird habitats through shade-grown coffee practices. Bird Friendly coffee is grown under a canopy of trees that provide habitat for migratory birds, contributing to the conservation of biodiversity.

For consumers, these certifications offer additional options for supporting sustainable coffee. By understanding the different certifications and what they represent, consumers can make informed choices that align with their values and contribute to a more sustainable coffee industry.

Chapter 5: Choosing Eco-Friendly Coffee

Reading Labels and Certifications

When shopping for coffee, reading labels and understanding certifications is key to making eco-friendly choices. Labels provide valuable information about the origin of the coffee, the farming practices used, and the certifications the coffee has received. However, with so many different labels and certifications on the market, it can be challenging to know which ones truly reflect sustainable practices.

Start by looking for certifications like Fair Trade, Rainforest Alliance, USDA Organic, and Bird Friendly, which indicate that the coffee was produced in a way that meets specific environmental and social standards. These certifications provide assurance that the coffee you're buying supports sustainable farming practices, protects the environment, and promotes social equity.

In addition to certifications, look for information on the label about the coffee's origin. Single-origin coffees, which come from a specific region or farm, often provide more transparency about the farming practices used. Many sustainable coffee brands also include information about their direct relationships with farmers, highlighting their commitment to fair trade and ethical sourcing.

By taking the time to read labels and understand certifications, you can make informed choices that support a more sustainable coffee industry and ensure that your coffee habit has a positive impact on the world.

Supporting Ethical Brands

Supporting ethical coffee brands is another important way to promote sustainability in the coffee industry. Many coffee companies are committed to sustainable practices, from sourcing beans from certified farms to investing in community development projects in coffee-growing regions. By choosing to buy from these brands, you can help drive demand for sustainably produced coffee and encourage more companies to adopt ethical practices.

When looking for ethical brands, consider their commitment to transparency and traceability. Ethical brands often provide detailed information about their supply chain, including the farms they source from and the conditions under which the coffee is produced. This transparency allows consumers to make more informed choices and ensures that the benefits of sustainable coffee production are shared equitably.

In addition to transparency, look for brands that prioritize quality and sustainability in every aspect of their business. This includes using eco-friendly packaging, reducing their carbon footprint, and supporting initiatives that benefit coffee-growing communities. By supporting these brands, you can enjoy high-quality coffee while contributing to a more sustainable and ethical coffee industry.

Local and Direct Trade Options

Buying local or direct trade coffee is another effective way to support sustainability. Local roasters often have close relationships with coffee farmers and are more likely to prioritize quality and sustainability in their sourcing practices. By buying from local roasters, you can support small businesses in your community while also ensuring that your coffee is ethically sourced.

Direct trade is a model of coffee sourcing that involves purchasing directly from coffee farmers, often bypassing traditional intermediaries. This approach allows farmers to receive a higher price for their coffee and fosters long-term relationships between farmers and buyers. Direct trade also promotes greater transparency in the supply chain, allowing consumers to know exactly where their coffee comes from and how it was produced.

By choosing local or direct trade coffee, you can support sustainable farming practices, ensure that farmers receive fair compensation, and enjoy fresher, higher-quality coffee. These options offer a way to connect more closely with the coffee you drink and contribute to a more sustainable and equitable coffee industry.

Chapter 6: Sustainable Brewing and Consumption

Eco-Friendly Brewing Methods

Sustainable coffee consumption doesn't end with purchasing eco-friendly beans—it extends to how you brew your coffee as well. Eco-friendly brewing methods focus on reducing waste, conserving energy, and minimizing the environmental impact of your coffee habit.

One of the most straightforward ways to make your brewing more sustainable is to use a manual brewing method, such as a French press, pour-over, or AeroPress. These methods require no electricity, reducing your energy consumption compared to electric coffee makers. Additionally, they often produce less waste, especially if you use a reusable metal or cloth filter instead of disposable paper filters.

Another eco-friendly brewing option is using a coffee maker with an energy-saving mode or one that automatically turns off after brewing. This helps reduce energy consumption and ensures that your coffee maker isn't running longer than necessary. If you prefer espresso, consider using a manual espresso maker, which requires no electricity and produces minimal waste.

Cold brew is another sustainable option, as it requires no heat and can be made in large batches, reducing the frequency of brewing. Additionally, cold brew can be stored in the refrigerator for several days, allowing you to enjoy coffee without repeatedly using energy to brew fresh batches.

Reducing Waste

Reducing waste is a crucial part of sustainable coffee consumption. One way to do this is by using reusable coffee cups and filters. Single-use coffee cups, pods, and filters contribute significantly to landfill waste, so switching to reusable alternatives can have a big impact. Many coffee shops also offer discounts for bringing your own cup, making it a win-win for both the environment and your wallet.

Composting coffee grounds is another effective way to reduce waste. Coffee grounds are rich in nitrogen, making them an excellent addition to your compost pile. They can also be used directly in your garden as a natural fertilizer or to deter pests. By composting your coffee grounds, you can help create a closed-loop system where waste is minimized and resources are reused.

Finally, consider buying coffee in bulk or choosing brands that use minimal or recyclable packaging. Reducing the amount of packaging you use helps cut down on waste and lowers the overall environmental impact of your coffee consumption.

Recycling and Composting

Recycling and composting play a vital role in reducing the environmental impact of coffee consumption. Many components of your coffee routine, from the packaging to the grounds, can be recycled or composted instead of ending up in a landfill.

Start by recycling coffee packaging whenever possible. Many coffee bags are now made from recyclable materials, though you may need to separate different components, such as plastic valves, before recycling. Some brands also offer take-back programs where you can return empty bags for recycling or reuse.

Composting your coffee grounds is another simple yet effective way to reduce waste. In addition to adding nutrients to your compost pile, coffee grounds can also be used in your garden to improve soil structure and deter pests. You can even compost used paper filters, provided they are unbleached and free of chemicals.

For coffee pods, look for brands that offer compostable or recyclable options. Many single-use pods are made from materials that are difficult to recycle, so switching to a more sustainable option can significantly reduce your waste.

By integrating recycling and composting into your coffee routine, you can minimize the environmental impact of your coffee consumption and contribute to a more sustainable lifestyle.

Chapter 7: The Social Impact of Sustainable Coffee

Empowering Farmers

Sustainable coffee production is not only about protecting the environment; it's also about empowering the people who grow our coffee. Smallholder farmers, who produce the majority of the world's coffee, often face significant challenges, including low prices, unstable markets, and limited access to resources. Sustainable practices help address these challenges by ensuring that farmers receive fair compensation and have the support they need to improve their livelihoods.

Fair Trade and other certification programs play a crucial role in empowering farmers by guaranteeing a minimum price for their coffee, providing access to credit, and promoting better working conditions. These programs also invest in community development projects, such as schools, healthcare facilities, and infrastructure improvements, which benefit not only the farmers but also their families and communities.

Direct trade relationships, where roasters buy directly from farmers, also help empower farmers by fostering long-term partnerships and providing higher prices for quality coffee. These relationships often include investments in training and resources to help farmers improve their farming practices and increase their yields.

By supporting sustainable coffee, consumers can help empower farmers, ensuring that they have the resources and opportunities to thrive.

Community Development

Sustainable coffee practices contribute to community development by promoting social equity and investing in local communities. Many certification programs require that a portion of the premiums paid for certified coffee be used to fund community projects, such as building schools, improving healthcare facilities, and providing clean water.

These investments have a lasting impact on coffee-growing communities, improving the quality of life for farmers and their families. Access to education, healthcare, and clean water not only benefits individual families but also strengthens the community as a whole, creating a more resilient and sustainable society.

Community development projects also often include training and education programs that help farmers adopt more sustainable practices, improve their yields, and increase their income. These programs empower farmers to take control of their future and contribute to the long-term sustainability of their communities.

By choosing coffee from companies and certifications that prioritize community development, consumers can play a role in supporting these vital projects and helping to create a brighter future for coffee-growing communities.

Gender Equality

Gender equality is an important aspect of social sustainability in coffee production. Women play a crucial role in the coffee industry, from farming and harvesting to processing and trading. However, they often face significant barriers, including limited access to resources, education, and decision-making opportunities.

Promoting gender equality in coffee production involves ensuring that women have equal access to resources, training, and opportunities for leadership. Certification programs and initiatives that focus on gender equality help address these issues by providing women with the tools and support they need to succeed in the coffee industry.

Empowering women in coffee production not only benefits the women themselves but also their families and communities. Research shows that when women have control over their income, they are more likely to invest in their children's education, healthcare, and nutrition, leading to better outcomes for the entire community.

By supporting initiatives that promote gender equality, consumers can help create a more inclusive and equitable coffee industry where everyone has the opportunity to thrive.

Chapter 8: The Future of Sustainable Coffee

Innovations in Coffee Production

The future of sustainable coffee lies in innovation and the continued development of new practices and technologies that reduce the environmental impact of coffee production while improving social and economic outcomes for farmers. One of the most promising areas of innovation is the development of new coffee varieties that are more resistant to climate change, pests, and diseases.

Climate change poses a significant threat to coffee production, with rising temperatures and shifting weather patterns making it more difficult to grow coffee in traditional regions. In response, researchers are working to develop new coffee varieties that can thrive in these changing conditions while still producing high-quality beans. These innovations not only help protect the future of coffee but also support the livelihoods of farmers who depend on coffee for their income.

Other innovations in sustainable coffee production include the use of technology to improve efficiency and reduce waste. For example, precision agriculture techniques, such as the use of drones and sensors, can help farmers monitor their crops more effectively and apply resources like water and fertilizer more precisely. This not only reduces the environmental impact of coffee farming but also helps farmers increase their yields and income.

As these innovations continue to develop, the future of sustainable coffee looks promising, with the potential to create a more resilient and sustainable coffee industry.

Challenges and Opportunities

While there are many exciting opportunities for the future of sustainable coffee, there are also significant challenges that need to be addressed. Climate change, in particular, poses a major threat to coffee production, with rising temperatures and changing weather patterns affecting the regions where coffee can be grown. As coffee-growing regions shift, there is a risk that traditional coffee-producing communities could be left behind, losing their livelihoods and cultural heritage.

Another challenge is the economic viability of sustainable coffee production. While sustainable practices often lead to higher quality coffee and better outcomes for farmers and the environment, they can also be more expensive and labor-intensive. Ensuring that these practices are economically viable for farmers, particularly smallholders, is crucial for the long-term success of sustainable coffee.

Despite these challenges, there are also significant opportunities for the future of sustainable coffee. The growing demand for sustainably produced coffee presents an opportunity for farmers to adopt more sustainable practices and access premium markets. Additionally, the increasing awareness of consumers about the environmental and social impact of their coffee choices is driving demand for more sustainable options.

By addressing these challenges and seizing these opportunities, the coffee industry can create a more sustainable and equitable future for coffee production.

Consumer Role

As consumers, we play a crucial role in shaping the future of sustainable coffee. Our choices—what coffee we buy, how we brew it, and how we dispose of waste—have a direct impact on the environment, the livelihoods of farmers, and the sustainability of the coffee industry.

By choosing to buy coffee that is certified by reputable organizations like Fair Trade, Rainforest Alliance, and USDA Organic, we can support sustainable farming practices that protect the environment and promote social equity. Supporting ethical brands and local roasters that prioritize sustainability also helps drive demand for sustainably produced coffee and encourages more companies to adopt ethical practices.

In addition to making sustainable choices when buying coffee, we can also reduce our environmental impact by adopting eco-friendly brewing methods, reducing waste, and composting coffee grounds. These small actions, when multiplied by millions of coffee drinkers around the world, can have a significant impact on the sustainability of the coffee industry.

Ultimately, the future of sustainable coffee depends on all of us—farmers, producers, companies, and consumers—working together to create a coffee industry that is not only profitable but also ethical, resilient, and sustainable for generations to come.

Conclusion

Recap of Key Points

Sustainable coffee is about more than just producing a good cup of coffee—it's about ensuring that the entire process, from bean to cup, is done in a way that protects the environment, supports the livelihoods of farmers, and creates a better future for all involved. Throughout this book, we've explored the various aspects of sustainable coffee, from the environmental impact of coffee production to the social and economic benefits of fair trade and ethical sourcing.

We've discussed the importance of understanding the journey from bean to cup, recognizing the environmental and social challenges at each stage of the process. We've also highlighted the role of certifications and labels in helping consumers make informed choices, and we've explored the impact of sustainable farming practices on the environment and farming communities.

By choosing to buy sustainably produced coffee, support ethical brands, and adopt eco-friendly brewing methods, we can all contribute to a more sustainable coffee industry. Our choices matter, and together, we can help create a future where coffee is not only delicious but also sustainable.

Call to Action

As you enjoy your next cup of coffee, take a moment to think about the journey it took to get to your cup. Consider the impact of your choices and how you can make a difference by supporting sustainable coffee. Whether it's choosing a certified organic coffee, supporting a local roaster, or composting your coffee grounds, every small action adds up.

The future of coffee is in our hands, and by making sustainable choices, we can help ensure that coffee continues to be enjoyed by generations to come. Let's work together to create a coffee industry that is fair, sustainable, and delicious.

Notes:

Notes:

Notes:

Notes:

Notes: